WATCHING THE PERSEIDS

WATCHING THE PERSEIDS

Peter Scupham

Oxford   New York
OXFORD UNIVERSITY PRESS
1990

Oxford University Press, Walton Street, Oxford OX2 6DP

Oxford New York Toronto
Delhi Bombay Calcutta Madras Karachi
Petaling Jaya Singapore Hong Kong Tokyo
Nairobi Dar es Salaam Cape Town
Melbourne Auckland

and associated companies in
Berlin Ibadan

Oxford is a trade mark of Oxford University Press

First published in Oxford Poets
as an Oxford University Press paperback 1990

British Library Cataloguing in Publication Data
Scupham, Peter, 1933–
Watching the perseids.—(Oxford poets).
I. Title
821.914
ISBN 0-19-282785-5

Library of Congress Cataloging in Publication Data
Data available

Typeset by Wyvern Typesetting Ltd
Printed in Great Britain by
J. W. Arrowsmith Ltd, Bristol

Acknowledgements

Some of these poems have appeared in *The Cambridge Review*, *Encounter*, *First and Always (Faber and Faber)*, *PN Review*, *Poetry Now (BBC2)*, *The Poetry Review*, and *Strawberry Fayre*.

Contents

The Good Thing

Will the good thing come, laired in its wet hedgerow,
Grown sick and shy in the rank smell of elder
And June's half-light, dangling its carapace
Over the less-than-wind, these loaded branches?
Sighing flower-heads, bushed inclinations,
Make their pale gestures to my outstretched hands
Scorched brown from holding out unchancy gifts.
What is it that these shades are fostering
In all their awkwardness, their nods and shakings
Pushing about this stillness of close air,
A mess of stony earth and broken feathers?
Will the good thing come to these unbudding fingers
Barking their skins against a nest of shadows?
Unreadable, the scribbling of the leaves,
And for all the wisdom they lay up in green
The trees can only answer yes, and no.

White Things, Blue Things

White things, blue things, their stalks tenuous,
The green pale where it moved to root-life;
Brought home, which was not their home,
Although they made it feel more like home
Than all that furniture: forests of dried sap
Creaking dry boughs where the lost wind
Sharpened to a draught cornering the room.

And, of course, though not asking to be,
Picked by the wayside; not asking anything
But dumb life, set in its green confusion,
Drawn always out of what lies hidden.
Plenty left to litter Spring and Summer,
Leaving each other about to be uncounted
Or seed themselves to sleep when time said so.

Crossing themselves in glasses, hung untended
Against a window-pane, out of their element,
Their gauzes drawn between us and the sun,
They look back out of memory and snapshots
Adding their lustres to 'Do you remember?'
Their heads calm, unshaken, when we ask
Our other question: 'Will they last in water?'

Paradise Gardens

I remember them inching gingerly into the past,
Testing their thoughts against the shine of clay:
A woman propped in a high-back chair, a prim sharp face,
Her dress the black shadow of her near-century—
The sand poised in my glass, hers running fast.

Old heads are heavy with dull wonders they cannot share,
The strength of root-life, tales told out of school.
They see the dead, warm and unwrinkled, walking
Down Paradise Gardens, up the Line, keeping still
To their careful patches of rain or sunlight, air

Which blackens roughly out. They know the closest names
Of those who tease them with sudden laughter, tears,
Forcing dreams with their passwords, tripping on birthdays;
We arrange the past years as neatly as cut flowers
But kisses burn and scald. There are live coals, flames

Before the exposure of ash, the short years before
We too must pass beyond the last visible ghost
To a time when all we wanted to say, or said,
Will be, like Etruscan or Linear A, quite lost,
As opaque as those banks of love on which we draw,

Which are buried so deeply behind our clouding faces.
The post is blankly angled by the lost tennis-lawn;
The invisible net catches nothing over and over.
There are faces adrift in all the window-panes
Mooning at rain, which is twisting their garden grasses

With tides of air which have no use for the dead
But to blow them back into graveyards and out again.
I stand, scraping green lichen from a tall family stone,
Allow cut names to shine, and all that I learn
Is how the signposts stood by roads down which I am led.

Familiars

You have grown terrible to me in your love,
Who did not ask to be forgotten, leave the lawn,
The careless signature, lit breathing-space
And courtesies of your good neighbourhood
For wind ruffling the stillness of those acres
Where you keep sullen house under spent flowers.
You are my dream-walkers, absent-minded ones
Pressing my head close to your fusty wardrobe,
Lost scents of flesh ageing their creature comforts.
I trust myself to your light-fingeredness,
Caught between kindly skin, the skeins of bone;
And, when I speak to you, my dear familiars,
I only take three simple words back home
Which wandered, lost, between your deaths and mine,
Chancing 'I love you'—which you knew already,
Turning from scribbled lawns and signatures
With half-smiles quite untroubled by such truths
As these words carry down my whispering gallery.
Oh, you slight ghosts, grown frail as all ghosts must,
More luminous, sustain me from your shadows
On these late errands, posting your dead letters,
Your small-change bright against my spendthrift hand.

A Red Ribbon

For my Aunt, Marjorie Wallis

The long passage is a climb of wet ice:
Cocytus, river of mourning, where all tears meet
And starry antechambers hold in place
Those whose eyes have fallen out of our light,
Wandering where memory yellows at the corners.

Shadowed in creamy white she lies in wait,
Touching and feather-brained as an old song,
This great girl: her hair fined to a plait
In a blossom of red ribbon, her traipsing tongue
Picking her brains, telling untidy lockers

That this is a prison, warded by sheets and bars,
To which, picked up stunned from her bungalow floor,
She has been sentenced for living beyond her years:
The best hotel—room-service. For here,
All are booked in; need make no reservations.

How good of me then to come for the last time,
And share in those happy nevers we must have had,
Whoever she is, whoever I was, or am,
Watching a red ribbon dancing over the bed,
The usual flowers draining their glass of water,

And how much she wants me to know, lost for a name,
Puzzling me out her son, her nephew, brother,
That she is much loved, lucky to be loved by them:
Those who have seen her so far, will see her further
Through ice, Cocytus, the sealing of all orifices.

The Master Builder

For my Uncle, Tommy Scupham

On leave, he must have slung his British Warm
Across this chair, hugged, kissed, and joshed them all—
That fire still warms the corners of the room.

His cap-badge showed the way the land could lie
For one more tommy coming home for good:
The Gunners' wheel above its *Ubique*.

He found the House that Jack built tumbling down
About their ears: the devil left to pay.
He picked things up; his ropes could take the strain.

He squared the family circle, found a skill
For throwing sky-hooks up against the clouds;
With blood for mortar raised the roof and wall

That set his brother and his sisters free.
They packed their cases, off to pack their brains
By ways that he had made. They made their way.

And where he grew, the little Market Town
Spun at his finger-tips: that tarnished wheel
Flashed out its golden spokes; became the sun.

A Monumental Mason. Near and wide
He played the jester, but they knew him king,
Bone-shaking off to Charlestons in his Ford,

Hanging one-handed, striking down a grin
From church-clock high: a monkey on a stick,
Or banging aces at some farmer's lawn.

He stood in his foundations, making clear
His rule, his measure. Children trooped the Yard,
The House that Jack built now set fair and square.

And still the family ghosts go whispering,
Whatever houses they inhabit now:
'You owe your brother Tommy everything.'

Teasing me still, he strolls the garden where
His old Victoria Plum nods in the wind,
And *Ubique*: his hands build everywhere.

Eldorado

Remembering Francis Norman, Bookseller

And so, no Saturday door to be angled through
    And find you again,
Puzzling about our latter-day quids and quidnuncs,
    Your Proust, Montaigne,
While Heath Street under a patchy cloudscape suffers
    A water-stain,

Or shuffling in, your keep-net a Sotheby's bag,
    (The catch indifferent today),
With quiz and chaffer and chat, with news of the ones
    That got away.
They settle to lighter dust on my patchwork shelves,
    Your Dryden, Gray.

A Professorial Chair, the spoils of Time
    Brown at your feet,
Babel: her grammars and guides in curling vellum,
    Collated, complete,
Long to be pickled in drowsy Academe's
    Cool winding sheet,

The learned Doctors, Johnson and Eustace Chesser,
    Ephemeris, Folio,
Cities of Dreadful Night and Beautiful Nonsense
    At criss-cross row—
And crumbs of learning tossed from your bread and cheese;
  In a sunset glow

Here then lay Eldorado, blocked in gold,
    Our Box of Delights,
The smoky torches of knowledge guttering down
    From Ancient Lights—
And a pinch of your salt still left for us to savour
    On Attic Nights.

A couple of rooms, all sects and centuries
    In broken run,
And your late courtesies, your text and gloss
    Victorian.
Now, the catalogue checked, and back to the printer:
    Finis, and Colophon.

The Christmas Moon

Cry for the wrinkled moon,
Pale face, box of tears;
Sharp-set on its line,
Pegged out by brittle stars.

Prop your dream-ladder; hang
Your head in miles of air
Frosting the topmost rung,
While the whole cirque d'hiver,

That feral zodiac
With blue and glittering eyes,
Fixes you in its lock,
Brings you to your knees.

Ask the cross-hatched night,
The brooches of cold flame,
That ghost in silver plate
Crossing your outstretched palm,

For this one night the marred
Changeling to be gone;
The stolen child restored,
Slipped back in your skin.

*

Something you asked for,
And it wasn't given:
Light from a lucky star,
Pennies from heaven?

The whole house lit
By grace and favour,
Doors on the out shut,
The bad dreams over?

Just what you didn't want:
Your face caught in the glass,
As innocent as paint,
Gift-wrapped on its loss.

*

The wishing moons fall:
A clutch of silver
Down a dark well.
The nights cloud over,

The hour-hands move,
Blurring orb and cusp:
Faces of love, of love—
Just out of grasp,

Each soft nimbus,
Unhealing bruise,
Setting your compass
Twirling its rose.

Travel by ancient light
Glanced from the dead
As it grows so late
On the beast road.

Lantern, dog, thorn-bush
Will see you home:
Silver dragged from ash
Spell out your proper name.

Levels

By shivering easels
The Old Dutchmen
Drew the days in,

Muttered to their sables
About clogs to clogs
In three generations,

How they stamp clay down
To the dull as ditchwater
Of life's quotidian.

Not enough blue here
To patch their trousers;
Just winter, water,

Finding their own levels
As the heart does,
As the leaves do,

Leaving a chink
For starved fingers
To keep time from freezing.

YOUNG GHOST

For my Mother, Dorothy Scupham, 1904–1987

1 Arriving

Whose are these faces clinging to the gate
That seem to have been left out in too much rain,
Blown-back leaves shimmered on a bent pane,
Which soften, work loose, fall? It is getting late

As I swing round on shoals of small stone
Crowded with damp air, skirmishing about
With blocks of shadow, this whole house wishing
It had been somewhere else. Here is where alone

Is growing visible, spreading its mesh wide,
Easing itself always between things left and taken.
The car is drenched with mud, dance-tunes shaken
Out of the past's throat as I drove where the trees died

Along the Roman road; from each tall going-down
The sap dried, a December night half-healing
The struck sets of bone, the growth-rings wheeling
From the young sapling to the full tree: crown

Crushed out, roots loose to the stars. I am there,
Where something fragile as blur on the warm windscreen
Hangs to old furniture, the once, the might-have-been
Of things. The faces rustle to the wind, asking me where

They must take themselves, being creatures of blossom,
    gleam,
Shifts of nothing much on brick, the interlace
Of black waxes, thickets of lost green: the wrong place
Gathers the ghosts of love into its own right dream.

2 Moving Round the House

Moving round the house, watching the shade
   Weaving loss into the Norfolk light
Blowing from rusted petals, closing the wide
   Places in afternoon which smells of night,
And hung with trees keeping their branches crossed—

I see things out of the corners of their eyes
   Glancing away from the huddle of the bed,
Staining the carpet, making their armistice,
   Drawing a late sun down from that overhead
Where God re-works his face from the cloud-scatterings,

Pulling the sheets, oh, slowly over the house
   Until that hugeness mixing its whites to grey
Coaxes the room to its own absent-mindedness—
   All the long nothings we have left to say
Lost in the dark that anchors the winter-garden.

3 The Fence

The garden had not much to say
Except that that was where stuff grew
By paths which ground themselves away.
Obedient to the names you knew,
It could not help those names to fit,
Although it made the best of it,

And putting on the usual show,
Brought all its flowers out of bed
In tetherings of to and fro;
In sun that brought things to a head.
Some other garden held your eyes;
This could not take you by surprise,

But studied your indifference
In blurs of white, or brown and green,
Beyond the dark, unyielding fence
That you had set so high between
The spaces you were moving to;
The garden only passing through.

4 *This Evening*

I watch you turning into memory,
Becoming something far too sharp, and clear.
You dwindle on the bed, and make a space
As small as Alice, waiting far from here

In rooms the size of photographs, on sands
By seas whose waters could not brim a cup;
Dressed in an inch or two of brilliant light:
Profound, and serious, and un-grownup.

As if you spend this evening trying on
The proper size to slip into a head,
Knowing how small the clothes are that must fit
That plain and simple hugeness: being dead.

5 Old Hands

Old hands, burnt out, the sap fallen: talons,
And a gold ring cutting closer to the bone—
The hands of children leaped across to meet them.

Always something left to do, undone,
Before the huge refusals of the world:
Those awkward women sitting in sad kitchens,

Fretting for stuff the years had dropped and broken.
Her green fingers worked best in the cold,
Dead-heading their autumnal gardens,

While her own border flourished its white spires,
Glooms where her stray cat, *Sawdust My Love*,
Crouched with bleak eyes, pads locked on tiny claws.

6 *Christmas 1987*

Once more I cover this nakedness which is my own,
Beyond shame, wasting, wanting; soon to be ash,
Pulling her nightdress down over the sharp bone
Which bites through the shocked absences of flesh.
    I hold viburnum up to her face; it is the Host,
    The offer of wax to wax—I am becoming a ghost

As I fuss for the damp sheets rucking across the bed,
My breath crossed upon hers when the choked sighs come
And spoons of morphine swim to the propped head.
It is time to die in this house which was never home.
    Rough winter flowers grow slack against the light;
    The days are another way of spelling the night

In this molten place where a silver photograph-frame
Runs like mercury, shifting its beads about,
Rocking her once again in my father's dream.
In the long-ago garden, holding the summer out,
    Her eyes laugh for his love: the ridiculous, wide
    Hat over dark hair, the white dress poured into shade,

This house burning away in my arms: the flowers,
The pictures loose with their huge and radiant trees,
Their cornfields crushed and rippling to faraway spires,
The cards blurred with angels and unpacked memories,
    A letter alert for the quick ply, the thresh of life—
    I am cut between glittering knife and glittering knife,

All honed by dangerous love. I drift through signs,
Defenceless, for there is nothing left here to defend
But the wrapped gifts helpless in their tinsel chains.
Somewhere Christ has been born; here, the promised end
    Thirsts for that promise kept: her one desire
    The pure trouble of fire, and what will become of fire.

7 At the Gate

By something about his air I knew he had seen me.
I had thought him left on the bed where a woman had died,
Lying in wait for his box and his gift-wrapped flowers,
The fire by which he had asked to be purified.

But there he was at the gate, juggling the sunshine,
Propping his cloudy windows against the view
Which kept the bricks and the grass in their usual places,
Leaving it up to his light to see things through,

As if those sharp-cut leaves, the restless traffic,
The blossoms bobbing about in their scuds of rain,
Were long breaths drawn and held in a foreign country:
A language cut on the cross, and nothing plain.

I knew the intransigent way he tugged at the world:
The rucked-up gravel, sun split by quartering trees.
He rubbed the skin of the day between his fingers;
Worked himself up into broken surfaces.

And this I should have foreseen; fostered, reminded
By what the dissolving landscape had come to prove,
Of how, by a different gate, from a different bedside,
He had broken the daylight in. His name was love.

8 Blessing

We are giggling about like girls in the Registry Office
As I hold her death in my hand. Can I spell its name
For the stunned typewriter, jolting its bones together,
The circumlocution of walls, where the nineteen-forties
Wobble their ghosts in blisters of cream and green,
Sigh in these box-files and the scraps of drugget?
Oh, what is hatched from these matches and despatches
When the mountainous Registrar takes a flower from his
     drawer
And swims like a seal through this wedding, where no
     impediment
Stands between bridegroom, all his astonishing hair
A wave of peroxide, tumbling to sharp lurex,
And bride, whipped like cream, scenting the flowers
Which twirl in the gloom and dazzle her nervous fingers?
The whole room buckles to life, swells with napery:
Handkerchiefs, tips of icebergs: a watch-chain's anchor
Sliding deep into billows of navy. Voices roll
Out of the broadest of acres; their brasses jingle
Under the crisp rosettes pinned against flesh—
We are flooded with Eastern light, and there in my hand
I hold my mother—a small, excited child
In her plaits, her buttoned boots, her Edwardian trimmings.
About us, the high parade of a Hiring Fair,
The plash of dirt-straw under the wedged beasts,
The huge teams dancing on their feathered feet,
Dancing away with love, as I consign her
Back to the files, back to the cradling dust,
With all these blessings, warmed by the fires of those
Who bless without knowing how, or whom, they bless.

9 Childish Things

I stand, my mother's weight upon my arm,
My clothes as serious as the future hid
Among the coffin-brasses, all this calm
Of lilies, granny's face beneath the lid.
     We move through scents, and she is at my side,
     Lips moving now: fast falls the eventide.

The house falls too: perched on a coldish spring
It waits in silence for the knacker's yard—
The clocks, the cups, the bits of everything
Dim out their lights; they too are dying hard,
     Losing that gloss I'd shimmered on their skins,
     Whirling away to strangers, widdershins,

For childish things grow dust, and go away;
That four-years' absence on a double-bed
Shrinks to the compass of an April day.
I had not thought how much goes with the dead,
     Nor yet how different my face could be.
     I pass the baked meats, and the funeral tea.

Again, the family faces gather round.
Suiting their darknesses to what they know
Must come: church flowers, cut sandwiches, the sound
Of huge words and small echoes. As we go
     I hear my father drily say: 'Your mother
     Would not have thought you out of character'—

These casual clothes—the stuff she could call *me*
If I had come in by a different gate
In different light. The bearers move, and she
Rides on their neutral arms. I take that weight
     Searching me out, and through. Past shine and blur
     Her mother's coffin-trestles wait for her.

10 Young Ghost

Oh, the young ghost, her long hair coursing
Down to her shoulders: dark hair, the heat of the day
Sunk deep under those tucks and scents, drowsing
At the neck's nape—she looks so far away,
Though love twinned in her eyes has slipped its blindfold—

And really she glances across to him for ever,
His shutter chocking the light back into the box,
Snapping the catch on a purse of unchanged silver.
Under those seven seals and the seven locks
She is safe now from growing with what is growing,

And safe, too, from dying with what is dying,
Though her solemn flowers unpick themselves from her
    hands,
The dress rustles to moth-wings; her sweet flesh fraying
Out into knots and wrinkles and low-tide sands.
The hat is only a basket for thoughtless dust.

And she stands there lost in a smile in a black garden:
A white quotation floating away from its book.
Will it be silver, gold, or the plain-truth leaden?
But the camera chirps like a cricket, dies—and look,
She floats away light as ash in its tiny casket.

11 Dancing Shoes

At Time's *Excuse-me*, how could you refuse
A Quickstep on his wind-up gramophone?
How long since you wore out your dancing shoes,

Or his vest-pocket Kodak framed the views
In which you never found yourself alone?
At times, excuse me, how could you refuse

His Roaring Twenties, stepping out in twos,
Love singing in his lightest baritone?
How long since you wore out your dancing shoes

And shut away that music, hid the clues
By tennis courts long rank and overgrown?
At times, excuse me, how could you refuse

To say his choice was just what you would choose,
Although he spun your fingers to the bone.
How long since you wore out your dancing shoes?

The Charleston and the Foxtrot and the Blues—
The records end in blur and monotone.
At Time's *Excuse-me*—how could you refuse?
How long since you wore out your dancing shoes?

12 1946

Our ghostly neighbours, walking their wounded rooms,
Dun her with stillbirth, madness: fierce dues,
And she must pay them out of an empty pocket.
Now, being chosen, how can she refuse?

Her black goodness blows to smuts on the wind;
Her dull sobbing sinks to the kitchen floor.
She fears cancer, being dissolved from us all
Into a nothing, a burnt offering, a for-ever-more

Of the cold stuff looked-out for on winter evenings
When we children press our noses to the pane,
Smelling dead light, worrying the road bare,
Lost on the trail of the late car and the last train.

She slams the door, slams her darks on the dark;
The house drops through the night like a black stone:
Her thoughts the jostle of rain in a loose cloud,
Her love bright as a child's fever, clenched as bone.

13 The Fire

The things they left unsaid
    Were less than air:
Shifts of an autumn sun
    From here to there.

There was the fire's low rush,
    That daring sibilance.
Its beads of steady light
    In the near distance

Drew all their secrets in,
    Each turn and glance—
Such hunger to consume
    Irrelevance.

As if the endless hour
    Of memories, names,
Was building a new dark
    From the old flames,

Feeding the chimney-stack,
    Whose sooty love
Posted what they had kept
    To things above:

Just a cold fold of tiles,
    Sugared by rain,
The spaces between the stars
    Clouded again.

14 The Finished Life

The finished life. And so it floats away
In rest and syncopation—an old song
As rough as breath, quotidian as the day,

To join the host of quick, completed things:
That clematis pinned back against the wall,
Those daisy-heads, adrift and thickening

Because the month has turned again to May.
I balance her upon my open page
As if she had become a holiday

As warm as sea expected to be cold,
Whose lights run out against a childish sun.
Here, she is nothing young, and nothing old,

But takes her chances from a sleight of mind
And, castled in her unrecorded hours,
Must lend herself to what the words can find:

A strip of ribbon burred across the skin,
Unstitched from choices worn to ravellings;
A spool of dark exposures winding in

Cups, conversations, an unfocused sky
Looking so clearly at the myriad grass—
And all we greet each time we say 'Goodbye'.

15 *The Stair*

I stop on the stair, look up to the lucky landing
      Where frosted sky
Aches apart on the early dead and the last.
Caught in the crook of the turn, my body bending
      To let you by,

I try once more to allay this trouble of faces;
      With double sight
Angling the dark for the truth of yesterday's ghost.
As the living slip to the sun, the dead to their houses
      Bricked out of night,

In a dusk retreat where much has been taken, given,
      I question air,
And under my feet that shuffle of ash and dust.
About you the tricksy shadows might have been woven
      From greyish hair.

As the hands from the dead-sea shore crowd an upstairs
  window
      Spliced by the moon,
I think of the places which keep the things I have lost,
And—tremor cordis—what I must do, undo,
      Knowing how soon

The ice in the bone will harden the outstretched fingers,
      How flesh, unmade,
Will scatter its tracings out on the lands of the past,
Into the haunts of the once-familiar strangers,
      My own ghost laid

Under this ladder of shades, the stair-rods ruling
      Their glancing bars
Across the climb of the night: still unconfessed,
Our share in that love which steadies the sky from falling,
      And lights the stars.

16 Birthday

Flesh touched, as if today
Something crosses a grave,
A poem, whose words carry
To anyone listening

Her sampler, silk house
Gardened by stitches;
Her box of Swiss music
Dancing on silverpoints.

Closed, she opens like the roses
Blown early in a dry summer
And lazy with girls' names.
She would have known them,

But speaks under the wind's breath
In twists of birdsong
Whose course she followed
Through woods, gardens,

Hidden in such green.
As June leafs over
The wish is a question:
'Happy Birthday?'

THE CHRISTMAS MIDNIGHT

I

The day slams down: the dark, the smell blocked thick;
I climb again up into the loaded car.
Thousands of miles are jumbled against the clock.
Black straps go tight on my electric chair.
The Sergeant-Major barks: 'I will run you in
So fast your little wheels won't touch the ground.
I'll spring you over the feathers, the bleeding skin,
The cliffs and the groan of sea at your land's end.
Shaking a forest of grass in your wind-machine
You'll be driven hard; at every straight and bend
Metal will duck from metal, the stars come clean.
As urgent and unpunctual as air
You will be somewhere. You will not be there.'

II

I do not think I am understood tonight
By things my head breeds language to understand:
Cold dials, green with a fishy light,
Cats' eyes glazing out from the swooped bend,
Those things which cross themselves, as if for luck,
And wave their hands to my vanishing curve of glass,
The wheel whose skin sticks to my fingers, the trick
Of shadows tossing me stuff I could hit or miss.
I play with loaded springs, my hands and feet
Dancing slow in close and musical air
As the engine drags and worries me out of sight,
The clutch slipped; time changing its gear,
Telling me how it will bring me safe out of harm:
Whinnying, trembling—my hand on the bonnet warm.

III

Tonight is black and white and go and come,
Is left and right—all swung antinomies.
The needle will not stick; it has to climb.
How many miles an hour to reach the skies?
I play with swings and roundabouts: a road
Where all these bits of driving, being driven,
Could make soft shadow, cloud, a being dead,
Floating for ever between Hell and Heaven.
I watch my ghosts whirl their excited hands;
They haul me through this pounding, heavy stuff
Whose cabs lean down: star-ships with clear-cut ends,
Their constellated lights a sliding graph.
I brush the road, light as a stolen kiss.
This is what I have come to. What is this?

IV

I come to myself, come home to chills of wind
In anchored trees, here, on this little hill.
My hands are shaking. It is at an end.
Light cracks and metals cool. An owl must call.
The car's thoughts settle, iron out its pain;
It snugs itself against the kindly frost,
Shrouds of light ice, ribs, webs of hardened rain,
Things ticking over, and things getting lost.
My cenotaph, my palace of black glass,
I lock your heart out with a silver key
As bright as moonshine. You shall be the place
Where I keep safe the absences of me.
In bed and open-eyed, with whips and goads
I drive you down my bright arterial roads.

V

I strap an image down. It bursts the seam.
My roof-rack thrums its cold Aeolian harp,
Chafing the stars, the wind—things as they come—
To burning elegiacs. Floating up
Between the night, the screen, his young-old face
A half-worked manuscript: intelligence
Wormed-out by maggot life, an empty house,
The Beardsley pull, the drink, his bright, intense
Low fires of loneliness, the phone that stretched
A spider-thread of thought to miles of pain,
The self-inflicted words, the silence. Wretch,
Find in his shadow your Samaritan
And clench your thought upon the girl who said
The living have the choice to be the dead.

VI

As dead as those who live: struck silences
Who dress themselves to meet themselves each day,
Floating like us in whispers, darknesses,
Lights dipped against the strangeness of the way.
The service-stations hospital the night,
Snapping the wind round coloured bits of tin;
Drugged by injection, music, chocolate,
We flicker back into our lives again
Dreaming of home: the dolls' house on the tip,
Warm chamberings for badger and for mole,
Honeymoon Lane—the shack on Smoky Top
Where someone's arms could try to hug us whole—
Scarecrows, lions, tin-men, all that crowd
Dancing round pot-holes on a witch-lit road.

VII

To trace the fault? With skill, with time, with love.
That alien murmur from the wear and tear,
Glossed-over, taped, roars into overdrive.
The lever frozen fast in second-gear,
What relay service comes to get you home,
After the crying, while the sobs go on?
Your plotted journey blown to Kingdom Come,
The more who pass, the more you are alone,
Pacing dull yards of tarmac, aching head
Full of cracked ice and pity for your sins:
Commissions and omissions. Down the wide
Careering freeway of the clean machines
The flash of orange beacon-lights—or blue—
Might steady climbing tumult; hunt you through.

VIII

Is to be young to wish the close of day,
Shiver at nightfall by the sway and hum
Holding your placard, letters rubbed away,
Watching the unrevealing headlights come,
So quickly go, leaving the dark unblessed?
Though the intelligent hitch-hiker's guide
Is out of print this year, and early frost
Picks at the fingers bumming for a ride,
Not only Spring-heeled Jack the Ripper stops
His reeking car; insinuates, cajoles,
With eyes as black as buttons, Bluebeard chops
And secret chambers for the cure of souls.
By luck, by fate, the good lifts come; the road
Smooths out a little underneath the load.

IX

The days are drawing in, are drawing down;
Their linen folds and blinds curtail the stars.
I start from cold; switch my ignition on:
That punctual red alert which disappears
Into its image. Advent. Stupefied
For some deliverance, I ease the brake
And slip into the habit of the road.
She will die soon. How long does dying take?
How many crossroads meet upon that bed
Where loaded years dismantle into love?
Pubs fairy-light their cypresses. Ahead,
The night is creeping skin; a shimmered glove
Whose blindfold sockets pull me to their ends:
Cradlings of bone, and layings-on of hands.

X

In this, the Christmas midnight of the year,
Viburnum flowers against the rain and cold.
I pitch my tent beneath its waxen star;
It is my myrrh, my frankincense, my gold.
In the huge shadows which are swarming her
She feels the scent cross miles I cannot cross;
Mothers me with a word. 'Lovely.' I hold the power
Of winter and its hidden promises.
There are no tears. Kneeling beside the bed,
We nurse her gently, under discipline.
I hold her far away, as far as God;
I wear her closely as a second skin.
We are as fluid as refusing air
That will not strip those sheets and pillows bare.

XI

The doctor stares, untongued. We shut the door.
Threads draw to breaking-point. I sleep alone.
He wakes me up into some darkness hour.
We look into the rags of time. 'She's gone.'
Hours he has spent beside her cooling flesh,
The *Holy Sonnets'* groundswell in his head,
Testing their weight, bare to each balm and lash.
There are some words to check. From out his shroud
Donne strikes lament and triumph: a dark bell
Whose golden tongue waits on the quick and dead.
I take his press of syllables, the spell
Of that communion, that wine and bread.
Truth at the well's foot. Our St Lucie's Night.
We sit in darkness reaching for the light.

XII

Always the Morello, its gouts darkening
Over a dream of white flowers, drifted sticks.
The end of a garden: a huge wall, light quickening
To slips riding on purples and soft blacks;
Always the Crab-Apple, and its painted gourds
Tumbled starrily on sweet-scented ground:
This down by a gate where I stand lost for words
Except those words I lost, which make no sound.
And quick on my lips the taste of bitter-sweets,
Of hard and soft found wanting, and found good,
Their sustenance weightless as these driven nights,
The bright fruit fresh and sour as my own blood.
I pluck them, pick them, press them against my tongue:
Savours which stay for ever, will not last long.

XIII

'Look in your mirror.' Look in it for luck.
It is packed with obfuscation, and with rain:
The past heaving about, its greasy slick,
Unharboured lights, things laid upon the line,
The sharp, unhooded eyes which fail to see
How close the dead are riding on my back,
Or how, beyond this bend, the land will lie
And how the road is about to break its neck
Where a string of lanterns keeps the narrow and straight.
His white beams locked on my dancing points of red,
I will lead him through a grimpen of mist and night;
I will guide him to something I might have understood,
Throttling the engine down to its beast-purr,
The moon as sharp as a hook in a velvet paw.

XIV

Ah, these blue undertakers in the rain,
Tender to wave me on or flag me down,
Comforting sick metal once again.
Sockets drip glass—how very much alone
These mortal houses with their painted skin,
Slewed on the verge of being what they were:
Between the sadness of what might have been,
The craning necks that cannot leave them there.
Banks carry flowers for them; a slow cortège
Winds awkwardly about these hats-off cones.
Can you see ghosts? There at the busy edge
Sits one released from his broken flesh and bones,
Drying his wings in sunlight: an ash look
Fixed on the pages of his judgment-book.

XV

Here's this, your ginger man, who does not know
I smell of death, but knows my engine's purr.
His simple tricks are magical as snow,
Boxing the compass by a sleight of paw.
He, too, has journeys he is bound to make,
Holds conversations with the night and rain:
Our speaking likeness that will never speak,
The great good cat, the looked-for thing, the sign,
A feral innocence with power to break
The human heart—supply his place again
With life that looks like love, like love can take
That puzzling something off that we call pain,
Ebbing away in furs which warm and sheathe
A quick-step pulse-beat and a sighing breath.

XVI

The scraped-out nerves of things will come to fledge
And even signposts learn to tell the way,
Old christening-robes of blossom dress the hedge.
Light could hold longer if it comes to May
When giddy people eat their cakes and cream—
But that's a distance yet. If it's inclined
To take me through a mirage to a dream,
A dream to the substantive and defined
I'd bless the road; would fit it like the glove
That fits my hand that fits the steering-wheel,
Keep a blue weather-eye on what's above
And out of reach; what I must bring to heel:
The smell of burning, and a ski-track skid,
A footing lost, a scorched and greasy road.

XVII

Oh, Sunday drivers, harnessed to your ease,
To blue-rinse bobbing head, upholstered lane,
May you, like flocks of sheep, most safely graze
The uncut verge, and Ambrose or Jack Payne
Put cream in coffee, pennies in small rain.
May water-lights from naked glass and chrome
Spring off to fuse with an embracing sun,
Your pad-nags turn their simple heads for home.
And you? My shadow, censor, *vrai semblable*,
Reined in by meadow-sweet or salt-licked sea,
Loose to the air my teeming courts of babel.
Clock up good miles. Let instruments agree,
And black horse, white horse, chariot, charioteer,
Drive four-in-hand the slip-roads of the year.

XVIII

Sometimes a light grows up to window-space,
A window-space to door. That light must know
How much it shares the nature of a star.
Would only fools or wise men think it so,
Stooping for miracles? Your scented tree
Has blossomed lustres all this Christmas-tide;
I picked a bunch of shadows, could not see
One candle burning for the flesh that died.
She holds me tight—as tight as I hold you,
Our winter-hands alert for quickening
Your garden, where Demeter dressed in dew
Holds out an urn to conjure with the Spring.
Tonight must hold us all, it is so wide—
Love born, love resurrected: crucified.

The Musical Box

The musical box has faded, as book-spines fade,
Taking the weight of the sun as the days come:
The key-hole out of key, the notes faltered,
The curry-comb gat-toothed on the spiked drum.

My gift, in lieu of that box in wartime Cambridge,
Going for a song, when songs were few, and dear.
In the half-lit house he opens it, posts his verses.
It is love: for her, for me. She is very near.

Memories? Let them rest in their quiet chambers,
The timeless gifts of time, forever ours,
But the one who is last to walk in the sunlit garden
Must watch for us both the play of the living powers,

The emerald lace-wing fly on the kitchen window,
The mating frogs with their hoarse and urgent voice,
The drift of long-tailed tits through the high box-elders
And the missel-thrush that sings for us both, Rejoice.

It is an age-old tune for a penny whistle,
Not for the pomps and stratagems of art:
It is a quiet tune that needs no playing
For us, who have the melody by heart.

He has two years left. When gifts return to the giver,
I take the box; find, read, and let it play,
This tune that needs no playing: a quiet chamber,
A coffin of silent sound and blurred inlay.

I crank it up, and under the ribboned glass
Let the jimp teeth flicker up on the rusting comb,
Sing of short turns about the sunlit garden,
Plantations, the Swanee River: old folks at home.

DYING

For my Father, John Scupham, 1904–1990

1 *November 24th, 1989*

My sister and brother-in-law think I should come
To see my father, who lies on his marriage-bed
Considering the seven ages of man,
Opening the windows on his Advent Calendar.
The books and pictures of his conversation
Place themselves calmly round about the room
For air to shelve and frame them.
Time slips its moorings. He eats little,
Sleeps to dream, and is pleased to see me.
His head is hoar-frost in a mild winter;
His eyes are smoke-blue. They do not waver.
It is time to mother and father his childhood,
Fetch and carry, ease him out of tangles.
Once he told me he wanted to be a poet
But decided that Yeats could do it better.
I would like to do justice to this mild, distinguished man
Who can handle seasons, flowers, the affairs of men,
Whose loosening mind keeps the Arnoldian touchstones
For sentinel on this, his longest journey.
He carries light. Dark will not hurt him greatly.
His veins are heavy plaits under a skin
Wearing its tenderness against the bone.
As muscle wastes, his tongue supports his weight,
Shaping, with all the clarity of reticence,
Sketches for the poems he never wrote
And would not have called poems if he had.
There is collusion. Each has his alibi.

2 'I have lived in a world which makes intelligible certain older worlds . . .'

'The things my mother made I'll never see again:
Potted beef, Lincolnshire fashion, flavoured with mace.
Food of the Gods. Melted butter on it. Marvellous.
Give all the pâtés in the world for it.

My mother was super-marvellous.
She came from a refined and educated home
And found herself married to a man who was neither.
She said: "Every one of these children
Is going as far up the educational ladder as they can go,"
And pushed them till they did.

By Scholarship, County Scholarship, County Major
    Scholarship,
We shall raise the wind—
Even if it comes out of savings—winter coats.

My father became an acute alcoholic.
Very serious indeed.
I spent morning after morning trailing him,
Deterring him from going into pubs.
Oh, it was a rum time.
He got money, and used to hide it in the garden,
But always knew how to bring half-pound slabs
Of butterscotch home for the children.
Broke it up with little hammers.

Mother's kind of heroism.
When she was quite old—her eighties—
Tommy arranged for her to go to classes,
University Extension Classes—poetry.
She wrote an account of a local bazaar
In heroic couplets.
She took vast pleasure in this.
She said: "I shall give this up.
Roger would understand me going to Bridge Parties;
He would not understand this."

I learnt two things:
To discount nine-tenths of what I read
About the ills of women in bygone days,
And that you could have a family at the mercy
Of mortal illness and bankruptcy,
And, given the right people and the right spirit,
It could be gorgeously happy.
And she was rewarded.
The children turned out diligent, and clever.

Her sisters all married professional men.
One of them said:
"Kate is the only one of us who's got nothing,
And the only one of us who has everything."

I know that the answer was the simple word "Love".'

3 'Oh dear, strange and long life is.'

'I would like to have been able to call myself a Victorian.
Of such small vanities . . .

I think I am that vacuum-cleaner,
And I sweep up gold-dust—very precious.

My ambition at the BBC—this sounds like vanity
But it has some truth in it—
My organization put out vast masses
Of simple biology, physics, geography, natural history—
My ambition was to be able to talk
To the producer of each specialism
In his own language and on his own wavelength.

Consequence was I met a lot of very surprising chaps:
Attlee, Herbert Morrison, Crossman—star turns—
Sutherland, Michael Ayrton . . .
The best minds of the day walked into my parlour.
Ronnie Knox—very agreeable.

"Man is but a reed, but he is a thinking reed."
That's Pascal.

I have a natural desire
To make myself more important and interesting
Than I really am. Of course, of course.

My ambition would have been to live a long life
With no events in it, and grow a great beard.

My aim now is to sit in a dressing-gown all day
Thinking disciplined, high thoughts.
I wish no part of the Contemporary Theatre.'

4 'Look in the next room . . .'

The books in the house are numbered
As carefully as the hairs on an old head.
Books behind books. There is a master-plan.
The building is the Celestial City.

'G. M. Young's dictum: read till you can hear people talking.
Not only what Wellington said to Creevey at breakfast
But Jacques le Bon to the grocer—
Behind the generals and the generalities.
Get me *The People's Armies*, by Richard Cobb—
Chap who lived in a fossilized watering-place.

If it were not for *The Children's Encyclopedia*
How should I know about *The King of the Golden River*?
How should I know about *Barbara Frietchie*,
Not to mention *Hiawatha*?'

He takes *La Rochefoucauld*, holds it upside down,
And reads the footnotes to his memory.

'I can do you the first hundred lines of *The Prelude*,
Or *A Toccata of Gallupi's*.

Fetch me Hallam's *Tennyson*—the second volume.
And if you see a book that's not worth reading,
Pass it to me, and I won't read it.

Even if it costs the earth
I want you to get me *Buried for Pleasure*,
Or *The Pleasure of Being Buried*
By Edmund Crispin.

As for those books from Waterstone's,
You could take them straight to Heaven and read them there.
I think the seating-arrangements would be better.'

5 December 26th

He knows that Christmas is somewhere round a corner.
We have put out his cards, read them to him,
Hung the odd picture with a twist of tinsel.
It is time for one of his favourite quotations:
'The feast of reason, and the flow of soul',
When the world perches on chairs, hands Chinese figs,
Sips the liqueurs he offers us by proxy.
It is Boxing Day, late afternoon. We bring the crib.
Virgin and Child suffer their nativity,
Doubled in his dressing-table mirror.
He has put an occasion together in his head,
And, at some far, unguessable cost, watches
A candle in a lantern, swinging softly.
'The Holly and the Ivy, when they are both full-grown . . .'
And, please, he would like Dorothy's favourite carol:
'He came all so still where his mother was . . .'
We eat, drink, and are merry. For company,
He nibbles air in front of a petit-four,
Asks and answers a little of the *Spectator* quiz.
For an hour, he is the Master of all Ceremonies,
Asking formal advice for 'friends who live next door' —
My sister's family — would we please suggest
A Place of Significance where they could spend
The three-weeks holiday he would like to give them.
There is a quotation. Something to do with
'Breathes there a man . . .' and 'plains of Marathon' —
I cannot see things clearly enough to find it,
And he would like each member of the company
To suggest a print, an etching he could buy,
Supposing he had that princely sum, a hundred pounds to
    spend.
It is not easy to put names to all the faces.
He is very tired now. He has managed it quite beautifully.
When the door shuts on voices, love and manners,
'Well', I say, 'you had your Christmas.'

'It was lovely.'

At three in the morning he calls for me,
His candle bright in its darkening lantern.
'Is it Christmas? Happy Christmas.'

'Happy Christmas.'

6 'Who do you say you are?'

'Life's difficult. How do you know you're not me?
You say you're Peter.
Do you want to persist in this claim?
You'll have a job to prove it,
Unless your documents are in very good order.

You were born in Bootle?
You couldn't say 'a fine city'
At the entrance to Bootle.

Anyway, who would want to be a poet?
Poets are rarely millionaires.

Tennyson had a dubious sense of his own identity,
And when he found himself alone, would murmur
'Far, far away, far, far away'.

I might deceive you about myself,
But I wouldn't deceive you about Tennyson.

I feel like a vacuum in an empty space.
Vacuums in empty space
Are very much at a loose end.
Now Blake wasn't at a loose end.

You know too much.
You claim you're Peter?
There are certain sorts of knowledge that are forbidden
And there's only one source they can come from,
And that's the Devil.

Are you the Devil?'

7 'The Scuphams are steadfast, inclined, you might say, to obstinacy . . .'

He demands a doctor, now, at two in the morning.
Threadbare, I lose control, shamefully, and say:
'Quite the little autocrat, aren't we?'—stumble to bed.
A little later he calls me softly.
'Would you like to hear the History of the Scuphams?
In epic verse. There are lyrics,
But they aren't all written yet.'

'That's much better. Yes, please.'

'I speak for the Scuphams, the Scuphams of Scupham Hall—
At least God knows if I really speak for them all.
But only the younger line, the older, trusting to belt and braces,
Having left no sons and heirs, and very few traces.

When the old confronted the new, face to face,
The Scuphams stepped right out of the Pilgrimage of Grace,
But I can tell you, in 1215,
The Scuphams were on the right side—you know what I mean.

Magna Carta, you fool!

The Scuphams never yielded to men nor beasts,
Neither to red-headed kings nor gay Francophile priests.

There's truth there—though I may exaggerate
The power of the Scuphams.

I have invented a deviant, weaker Scupham line.
Soon I shall come to Darwin, Newman and Pope Pius the
    Tenth—
When you get the intransigent stuff.
It is the business of the Scuphams
To stand up against French free-thinkers.

Scuphams show common sense, and a general scepticism
About foreigners and their intentions.

But I fear some of the younger Scuphams
Are showing traces of the *Heneage Beard*—
As long as your whole body.'

As long as an unfinished epic.

8 'What time is it?'

'Twenty to five? It can't be.
Do you mean twenty to five yesterday?'

The hours are pillows, propping us together.
I would call it night, the huge small hours.
I enter his space-time continuum.

'For God's sake. This is *urgent*.'

We must work hard, form a committee of absence
To sort out the affairs of Zing and Bing.
How can these villages get their washing done?
Land must be purchased, a machine installed,
Power-lines laid, teachers and clerics assembled.
On Sundays the washing can be blessed out to dry.

He restores the King, the Red Kite, in Wales,
Foils the egg-and-feather gangs,
Brings the lost thing home again.

On Smith's farm he works the harvest—
Six sheaves to a stook—
'Long, tall bones and magpies.'

And, waking on the floor in a twist of bedclothes,
Falls 'into the hands of a remote Scottish Clan
Of unbelievable barbarity, who fed me on porridge.'
We struggle with the dead weight of the living.

'I am all right, aren't I?
I have woken up in my own world,
Not in some nightmare system of pipes and sewers?

What have I been?
I thought I was some sort of animal in the night.
I want to wake up and be a man now.

You say it's ten,
But by the amount of experience that's gone into it
It must be at least half-past four.'

9 *January 10th, 1990*

For six weeks we have listened to our whispers,
Read signs and omens, taken our parts
In this play of seven ages. 'I've no doubt my nurses
Are good men and women, which only goes to show
Good men and women can be very tedious.'
He has lived beyond the possibilities of living,
His body sewn together by threads of thought,
The flesh pale, scarred, flimsy as a chicken-carcass.
He cannot speak. His eyes are constant, open.
The Doctor is interested. 'Uraemic frosting.
I would like my student to see this.'
In the night I read to him from *The Prelude*,
Celebrate a Midnight Mass for him
And the communion of his faithful ghosts:
The country child, damming the beck, skating,
At home in Legsby or in Willingham Woods,
Prospecting for gold-dust hidden in small flowers,
The Scholarship Boy, the Cambridge double-first,
The tennis-player with the kicking serve . . .

'Dust as we are, the immortal spirit grows
Like harmony in music; there is a dark
Inscrutable workmanship that reconciles
Discordant elements, makes them cling together
In one society.'

At three o'clock in the afternoon, he dies.

Cause of death
　　1a Uraemia
　　1b Carcinoma duodenum
　　1c Diabetes
　　1d Myocardial infarct

'We live by admiration, hope and love—
As someone said—it must have been Wordsworth.'

I glance to where he is quizzing his *Times* obituary.
Quietly, we continue our conversation.

Clearing

Four years ago, the house was waiting for them,
As cool and as convenient as light,
And each had eighty years to pack, unpack there,
　　To get things right,

As right as trivets, blackbirds, briar-roses,
Millennial stones and letters from old friends,
The winnowed books in which the past had printed
　　Beginnings, ends.

And from the sunroom window, looking inwards,
An eighteenth-century landscape on the wall
Showed trees whose leaves grew calm by a brown gradation,
　　But could not fall.

The days drew in; the days grew longer, deeper,
Made living-room for the tiresome, loving dead
Who shared their dreams, their food; who told them stories,
　　Left things unsaid.

The difficult chapters come to a kind of end.
The books rewrite themselves as fiction. Things
Fall easily into our hands as if they believed them:
　　Withdrawals, offerings,

As the dark stuff in its album strips away,
Leaving the house ghost-naked, white as a sheet—
The convenient light pausing at wall and window
　　For its heart to beat.

Watching the Perseids: Remembering the Dead

The Perseids go riding softly down:
Hair-streak moths, brushing with faint wings
This audience of stars with sharp, young faces,
Staring our eyes out with such charming brilliance:
Life, set in its ways and constellations,
Which knows its magnitude, its name and status.

These, though, are whispered ones, looked for in August
Or when we trip on dead and dying birthdays,
Drinking a quiet toast at some green Christmas
To those, who, fallen from space and height
No longer reach us with their smoky fingers
Or touch this sheet of water under no moon.

They are the comet's tail we all must pass through
Dreamed out into a trail of Jack O'Lanterns,
A shattered windscreen on the road to nowhere.
We stand in this late dark-room, watch the Master
Swing his light-pencil, tentative yet certain,
As if calligraphy could tease out meaning,

And, between a huge water, huger sky,
Glimmers of something on the jimp horizon,
There might be pictures, might be conversations.
We wait for last words, ease the rites of passage,
The cold night hung in chains about our questions,
Our black ark swinging lightly to its mooring.

Calendarium

I saw the cloudy genie burst his wine-skin
And blab large promises. On the dead ground
New gifts lay wrapped: a calendar of days
Unspent and unremarked. The wind idled.
At last I turned, and saw beyond the door
Resolutions thaw in the brief sunlight.
'I'll cook your goose.' The goose was cooked and eaten.
Under the Christmas trees I watched the children
Search for a brief kindness in the snow.

FEBRUARIUS

Fur sits by the hearth, yawning a red yawn,
Easing his claws. The year has turned,
But the dull acres lie all senseless,
Rain, dark and rain guessing at their contours.
Undo this button, set the chair closer
And thorns under the pot: a fool's laughter
Raising the roof-tree to the clouded stars.
It is a comfort to write long letters,
Urge frozen fingers to a little piece-work,
Stare at the nothings bobbing in the fire.

MARTIUS

Much can be seeded from a stiff yard-broom,
A duster shaken from an upstairs window.
Room now for the look of things to clear,
The pliant heart to suffer some excision.
It is medicinal to graft and prune
Under these wayward gusts, flecked by birds.
Set your house in order. Make a start.

APRILIS

All that you know, all that you thought you knew
Ploughs under as the share deals it.
Resurrection is in the air, the sky loose
In quick clouds, bold Easter wings.
Luck lies in a horseshoe; at the field's edge
I picked one out of shining clay, iron
Struck from a sweating team the years had harnessed.

MAIUS

Meet on the field-path, at the kissing-gate,
And read in love's fine print, still dancing
Its old measures in this press of leaves.
Untie the daisies; your sharp nails will bleed them,
Slit the pale stalks to die against her throat.

IUNIUS

In this cloudless weather, where reflections hang
Under the substance of their woods and cities.
Nothing is sweeter. Make hay while the sun shines.
In the white spaces of the longest day,
Urgent against the shades as yet undarkened,
Siren voices plague the drying grass.

IULIUS

In between moving to and moving from,
Unfolding green, disclosing darker green,
Lie deep fields, cornered by the eye of childhood.
In between moving up and moving down
Uncut plaits of wheat bend to the land's lie.
Space between then and now. Sleep, and good dreams.

AUGUSTUS

As you sow, so shall you reap. The bags packed,
Umbers and gold swollen between the purse-strings,
Getaway cars nose on a hot scent.
Under striped canvas the patrons gather,
Staring at blue, incorrigible seas.
The stubble burns a hole in summer's pocket;
Upon the baked crust of their world, the mice
Scatter their ashes to the harvest moon.

SEPTEMBER

Sun lingers out the senses; shawls of light
Ease the eye into fullness, and late suns
Polish our gourds. There, at the orchard gate,
The God deals out the first-fruits of the season,
Equates desire with his huge performance.
Much promise lies in apples, and in flagons.
Blue, though, is the colour of enchantment;
Enter the hooded skyline, leave the windfall
Ravished and puckered on the cooling lawn.

OCTOBER

Off with these lendings. An Indian summer
Cracks every seed-head under rising wind.
Time now to cauterize by fire
Outscourings from crooked corners:
Barren leaves, the sack of libraries
Enfolding in their Dead Sea scrolls
Runes whose ancient salt has lost its savour.

NOVEMBER

Night and the laurels hang their sad shields
Over parades of gravel by the lych gate.
Victory and defeat are out to grass,
Each known unto God by name and number:
Muck upon mould, caught between wind and water.
Blind scatterings, poppies of oblivion,
Ember days, uncalendared: the mists down further.
Remember. But what shall we remember?

DECEMBER

Daylight shortens; your anticipation
Eases the candle-flame towards its end.
Cassiopeia swings above the fir-trees,
Embraces night. Between such diagrams
Manger and child float down the centuries,
Brilliants nursed on soft cloth. Time is now:
Each life an altar where the Mystery
Repeats its dream of gift and sacrifice.

St Madron's Well

They might have crooked their fingers, turned things over,
Paring the fleshed moment down to bone.
In the silence of their silence something hung
As hopeful as an unglassed crescent moon,
As rough and brilliant as lost love, love lost
Where twigs cracked cold wishbones underfoot.

It was, perhaps, the hope to find a place
Where moonbeams could be lightly packed in jars,
And thought, half-thoughts, blurring to images,
Could rest their small-change on the shelving darkness.
Here, at the Saint's Well, where the strips of rag
Pinned such brave hopes to briars and unfledged elders,

They saw these things, and kept their own counsel,
As if each wish had not been wished before
By those, who, dead and human, only saw
Spring trying to prove that something could come true
In makeshift green; the salt of centuries
Thrown over their left shoulders by the wind.
The Saint standing in his ruined Baptistry
Turned towards them in a dress of leaves,
His hands, bleached to sea-air, blessing them
In the brown habit of his crinkled water,
And cold light crossed itself in holes and corners
As if there was something one could take for granted.

Last Offices

Smoked glass flares in the sun: a host of wafers
Posing a juggler's waterfall of cards,
A palace where the functionaries of process
Encode their black or beatific vision
And liverish house plants infiltrate slow tendrils
Between rolled shirtsleeves, scents and after-shave.

The untidy street works up to semaphore,
Feeding the tilted panes with talk of clouds
And silent traffic cutting the light's corners.
Glass runs like tears, and holds in its suspension
Each stroller and his object of desire.
They blossom to a cool millefiori,

As if the sky, stonewalling us with blue,
Was only something to prop up our windows.
Here, a Christmas hexagon, sealing snow
And every leafless garden of the heart,
Dies through half a century of frost-flowers.
Your nail scratches the ice to a white powder

And there are your sickroom ghosts; the hot glass
Waves to itself in a swirl of yellow curtains,
Leaves brush the sky as bedclothes brush the flesh
Beyond an endurance which must be endured.
The street walkers are lost in a crowd of rooks
Flocking their ash wings down to the falling elms,

And someone, pausing, a something in the city
Is crossed with a gardener and a slatted barrow
Stashed with brown leaves. A cat, as dead as Egypt,
Washes its face, primps, summering itself
Where faces ripen like a crowd of apples
Hung on the past, the far Hesperides.

Last Offices: a Grand Messe des Morts.
The stained glass burnishes its road to Calvary
And the low plain-song from lost terminals
Asks you its fitful questions. Search, recall
Your own curriculum vitae passing through
These rainbow filters, palindromes of light,

Before the nights draw in, this magic lantern
Projects itself as pure geometry,
And secretaries of dream redress the files,
Puzzle over petals, minutes of lost meetings
Or lift their heads from scents of grasses, roses,
To new moons rising, cold and thin as luck.

OXFORD POETS

Fleur Adcock

James Berry

Edward Kamau Brathwaite

Joseph Brodsky

Basil Bunting

W. H. Davies

Michael Donaghy

Keith Douglas

D. J. Enright

Roy Fisher

David Gascoyne

Ivor Gurney

David Harsent

Anthony Hecht

Zbigniew Herbert

Thomas Kinsella

Brad Leithauser

Derek Mahon

Medbh McGuckian

James Merrill

Peter Porter

Craig Raine

Christopher Reid

Stephen Romer

Carole Satyamurti

Peter Scupham

Penelope Shuttle

Louis Simpson

Anne Stevenson

George Szirtes

Grete Tartler

Anthony Thwaite

Edward Thomas

Charles Tomlinson

Chris Wallace-Crabbe

Hugo Williams